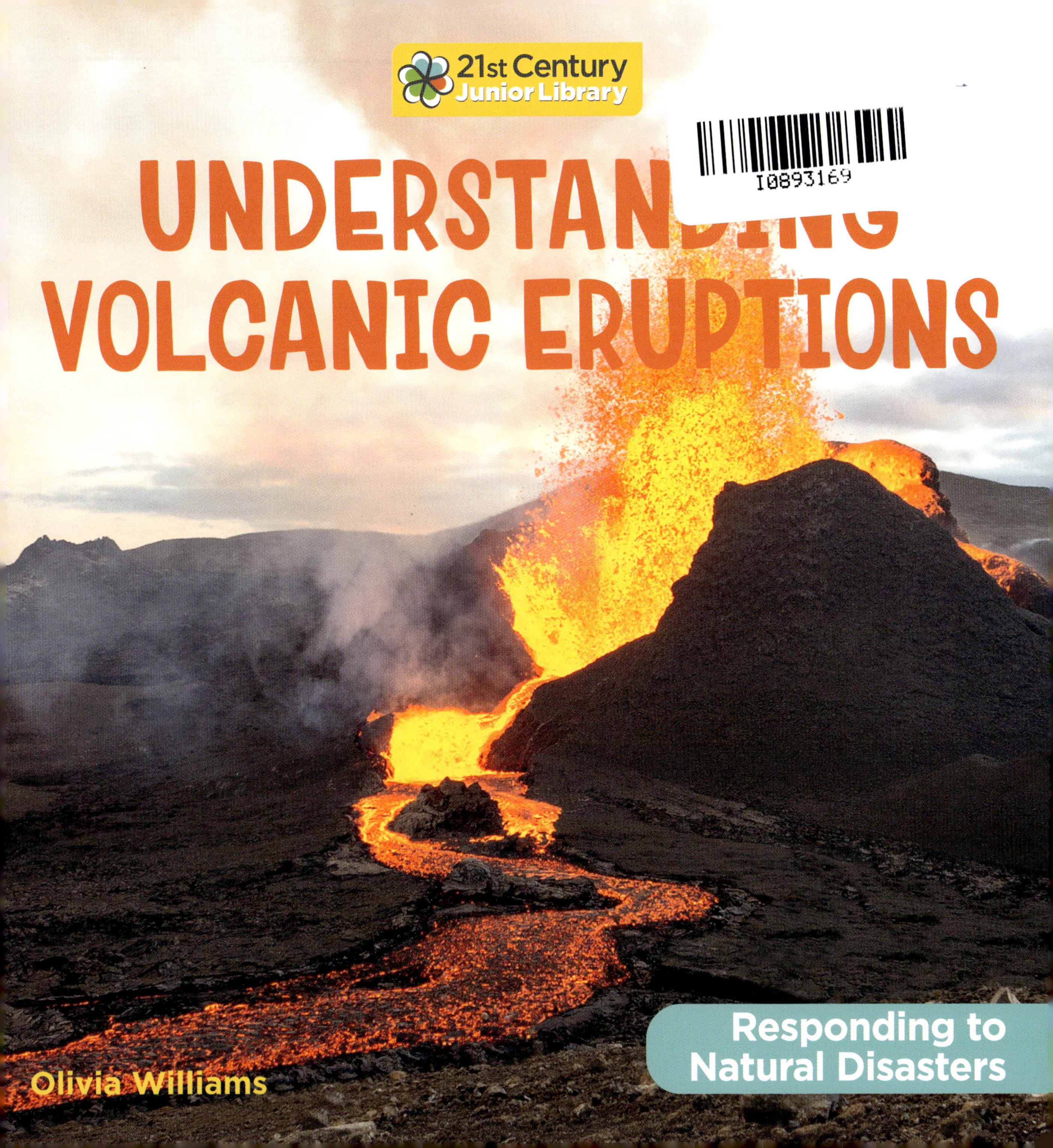

21st Century Junior Library

UNDERSTANDING VOLCANIC ERUPTIONS

Responding to Natural Disasters

Olivia Williams

Published in the United States of America by:

Cherry Lake Press
2395 South Huron Parkway, Suite 200, Ann Arbor, Michigan 48104
www.cherrylakepress.com

Reading Adviser: Beth Walker Gambro, MS, Ed., Reading Consultant, Yorkville, IL
Photo Credits: © Thorir Ingvarsson/Shutterstock.com, cover, 1; © Fotos593/Shutterstock.com, 5; © cristan pago74/Shutterstock.com, 6; © Yvonne Baur/Shutterstock.com, 7; © Larry Johnson/flickr, 9; © Pedro Carrilho/Shutterstock.com, 10, 11; © Olivier A. A. Vandeginste/Shutterstock.com, 12 [left]; © Anders Peter Photography/Shutterstock.com, 12 [right]; © tunasalmon/Shutterstock.com, 13 [top]; © Roberto Destarac Photo/Shutterstock.com, 13 [bottom, left]; © Sad Agus/Shutterstock.com, 13 [bottom, right]; © Francisco Sandoval Guate/Shutterstock.com, 15 [top]; © Paulo Miguel Costa/Shutterstock.com, 15 [bottom]; © RaffyT/Shutterstock.com, 16; © BMJ/Shutterstock.com, 19; © Budkov Denis/Shutterstock.com, 20; © BlueRingMedia/Shutterstock.com, 21

Cherry Lake Press is an imprint of Cherry Lake Publishing Group.

Library of Congress Cataloging-in-Publication Data
Names: Williams, Olivia (Writer of children's books), author.
Title: Understanding volcanic eruptions / Olivia Williams.
Description: Ann Arbor, MI : Cherry Lake Publishing, [2022] | Series: Responding to natural disasters | Audience: Grades 2-3
Identifiers: LCCN 2021043063 | ISBN 9781534199194 (hardcover) | ISBN 9781668900338 (paperback) | ISBN 9781668906095 (ebook) | ISBN 9781668901779 (pdf)
Subjects: LCSH: Volcanic eruptions—Juvenile literature.
Classification: LCC QE522 .W55 2022 | DDC 363.34/95—dc23/eng/20211005
LC record available at https://lccn.loc.gov/2021043063

Cherry Lake Press would like to acknowledge the work of the Partnership for 21st Century Learning, a Network of Battelle for Kids. Please visit http://www.battelleforkids.org/networks/p21 for more information.

Printed in the United States of America
Corporate Graphics

ABOUT THE AUTHOR

Dr. Olivia Williams is from mid-Wales in the United Kingdom. She has traveled to many countries to help people after natural disasters. She has felt earthquakes and aftershocks, and seen damage from floods, typhoons, and wildfires. The natural environment can be harsh but also beautiful. Olivia loves to hike and run in the mountains and photograph the incredible things she sees.

Cherry Lake Press

CONTENTS

WHAT ARE VOLCANIC ERUPTIONS?

Earth has four layers: the inner core, the outer core, the mantle, and the crust. The crust is the layer that we stand on. Sometimes there are cracks or **vents** in the crust. **Magma** from Earth's mantle pushes up through the vent. Magma is an extremely hot material that forms under Earth's surface. When magma escapes through a vent in the crust, we call this an **eruption**. When magma breaks through Earth's crust, scientists call it **lava**. "Volcano" is the

The word *volcano* comes from "Vulcan." Vulcan is the name of the Roman god of fire.

name we give to this process. It is also the name we give to the geographic mass created after an eruption.

Some volcanoes look like tall mountains. Others are short and wide. Magma from some volcanoes shoots up into the air. Lava from others slowly pours out. Magma and lava harden when they touch cooler air. As these

A volcano is still considered active if it has erupted within the last 10,000 years.

Magma and lava are liquid rock.

substances cool, they turn into rock. The rock creates the sides of a volcano. The sides may be tall and steep. Sometimes the sides are so steep that people can volcano surf down them!

Some volcanoes are active. Others are **dormant**. The highest active volcano in the world is Ojos del Salado. It is 22,615 feet (6,893 meters) high.

It's in the Andes Mountains in South America. South America is also home to the shortest volcano on Earth, the Pali-Aike volcanic field. This area is so flat and rocky that it looks like the surface of the moon!

Some volcanoes even erupt underwater. They are called submarine volcanoes. Volcanic eruptions can occur over a long period of time. Yasur is a volcano in Vanuatu in the South Pacific Ocean. It has been erupting for nearly 250 years and is still active!

Make a Guess!

Because Ojos del Salado is the world's highest volcano, tourists like to go there. Travel companies provide tourist tours around the volcano. What do you think it would be like to live somewhere that relies on tourism? What do you think the benefits and challenges would be for locals?

The largest active volcano is
Mauna Loa, located in Hawaii.

HOW VOLCANIC ERUPTIONS AFFECT THE PLANET AND PEOPLE

Crater Lake in Oregon is the deepest lake in the United States. It is 1,943 feet (592 m) deep. It is a dormant volcano crater that is filled with water. Native peoples watched it form 7,700 years ago. Volcanoes have formed many of our cultural **landmarks**. They have made some countries popular places to visit. Mount Fuji in Japan, Mount Etna in Italy, and Krakatoa in Indonesia

are all famous volcanoes. Eyjafjallajökull (AY-yah-fyah-lah-YOH-kuul), also known as e15, became internationally famous in 2010 when it suddenly erupted. It grounded airplanes. This means that they weren't allowed to fly. This was the largest air-traffic shutdown since World War II!

Vesuvius is a volcano in Pompeii, Italy. It is famous for its eruption in 79 AD. When it erupted, rock and ash

Around the world, there are about 1,500 active volcanoes! Many can be found in an area called the "Ring of Fire" in the Pacific Basin.

shot 21 miles (34 kilometers) into the air. More than 1,000 people died. The ash turned people and animals to stone. Today 2.5 million tourists visit Pompeii every year. They want to see what life was like for people in 79 AD. They also want to see what the volcano did to the town and its people.

Many active volcanoes are dangerous. But they can also help people. They create some of the most fertile soil in the world. For many years, people have lived near

Ask Questions!

In 2018, a volcano called Volcán de Fuego in Guatemala erupted. It killed 165 people and affected 1.7 million others. Guatemalan citizens said that the government was too slow to tell people to evacuate. What kind of things do you think the government had to consider? Do you think deciding to evacuate people is a difficult decision and why?

volcanoes to harness the **nutrients** in the soil. Nutrients make plants grow strong and healthy. Healthy plants provide lots of food for us to eat. South American countries like Colombia, Ecuador, and Guatemala grow coffee beans. Coffee growers rely on the Pacific Ring of Fire. Three-quarters of the world's most active and dormant volcanoes can be found there. This means that the soil there is excellent for growing food.

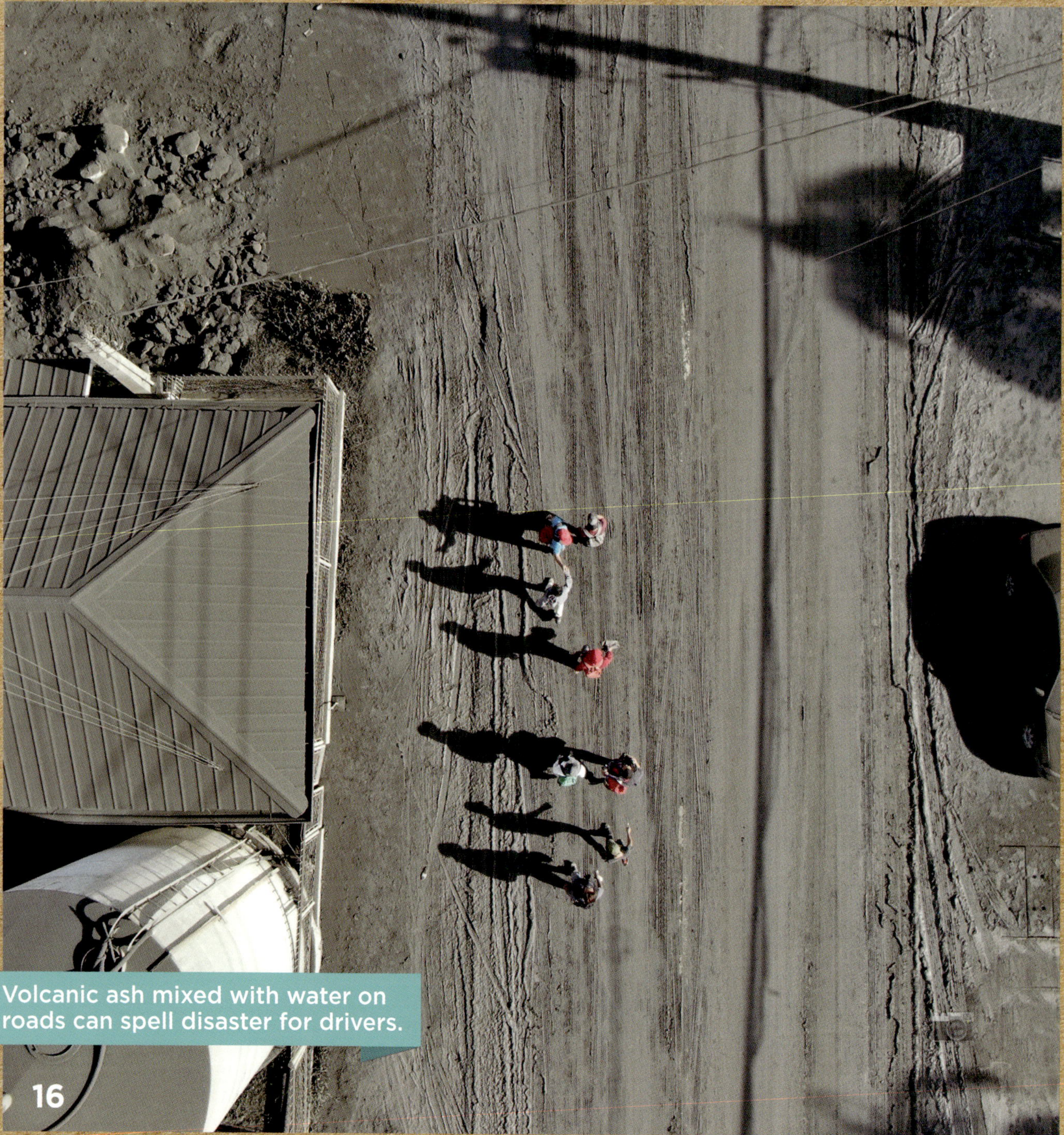

Volcanic ash mixed with water on roads can spell disaster for drivers.

PROTECTING AGAINST AND RECOVERING FROM VOLCANIC ERUPTIONS

Hundreds of years ago, volcanoes erupted with little warning. Now scientists use special tools to measure the activity of a volcano. Volcano scientists are called **volcanologists**. Volcanologists can tell governments when a volcano is likely to erupt. Governments can issue warnings telling people how to protect themselves. They may suggest things like wearing a dust mask, taking shelter, and drinking bottled water.

Recovering from a volcanic eruption isn't easy. Volcanoes leave ash **debris** everywhere and destroy people's homes. Volcanoes can be aggressive and scary. People are often left traumatized.

Think!

Think about the last time you spoke to someone about a problem. Can you name the feelings that you experienced? Imagine a friend has a problem. Name two things that you would do to make them feel comfortable talking to you about it.

It can take as long as 130 days for flowing lava to cool down to a solid state.

While volcanoes can quickly form new islands, they can also just as quickly destroy them.

For many people, fixing their homes isn't the only task. Sometimes people need to heal themselves too. Speaking with friends and family can help.

Create!

Communicating how you feel with a trusted friend or adult can help reduce stress. It can also help strengthen relationships and increase happiness. Create a "communication chain." Check in with a friend or family member. Ask them how they're feeling.

ACTIVITY

Make a Volcano

Learn the ins and outs of a volcano by creating different models! Use the following website to download and print four different volcanos based on active and extinct volcanoes from around the world.

Materials Needed:

- Printer
- Scissors
- Tape or glue

https://www.bgs.ac.uk/discovering-geology/maps-and-resources/earth-hazards/volcano-models

Parts of a Volcano

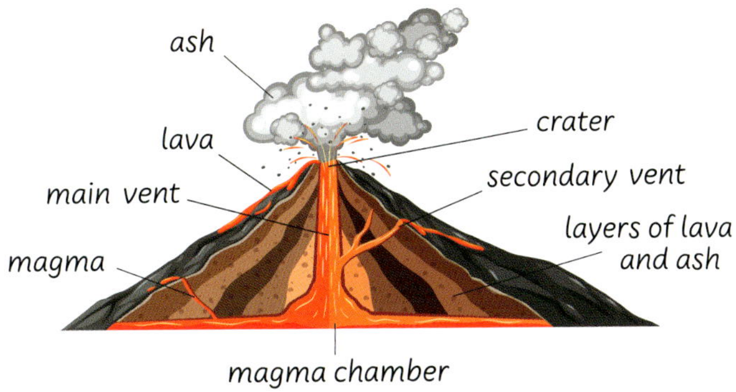

GLOSSARY

crater **(KRAY-tuhr)** a large bowl-shaped dip in the ground often found at the top of a volcano

debris **(du-BREE)** scattered pieces of broken rock or trash

dormant **(DOR-muhnt)** when a volcano or animal is alive but so inactive it is like it's asleep

eruption **(ih-RUP-shuhn)** when something such as magma shoots out suddenly and forcefully

evacuate **(ih-VAH-kyoo-wayt)** to remove people from a place to take them somewhere safer

landmarks **(LAND-marks)** things in the landscape that are familiar and help people figure out their location

lava **(LAH-vuh)** hot liquid rock that exists above the ground

magma **(MAG-muh)** hot liquid rock that exists below the ground

nutrients **(NOO-tree-uhnts)** substances that provide the needed minerals to grow and survive

volcanologists **(vahl-kuh-NAH-luh-jists)** scientists who study volcanoes

vents **(VENTZ)** openings in Earth's crust where magma can escape

FIND OUT MORE

Books

Fox, Janet. *Volcano Dreams: A Story of Yellowstone.* Berkeley, CA: Web of Life Children's Books, 2018.

Volcano & Earthquake. New York, NY: DK Publishing, 2022.

Website

Close to Home: Living and Farming with Volcanic Ash

https://youtu.be/Czzibq2LBbQ

Listen to South Americans describe what it's like to live near an active volcano.

INDEX